CHAPTER 2

CIRCUIT PROTECTION DEVICES

LEARNING OBJECTIVES

Upon completion of this chapter you will be able to:

1. State the reasons circuit protection is needed and three conditions requiring circuit protection.

2. Define a direct short, an excessive current condition, and an excessive heat condition.

3. State the way in which circuit protection devices are connected in a circuit.

4. Identify two types of circuit protection devices and label the schematic symbols for each type.

5. Identify a plug-type and a cartridge-type fuse (open and not open) from illustrations.

6. List the three characteristics by which fuses are rated and state the meaning of each rating. Identify a plug-type and a cartridge-type fuse (open and not open) from illustrations.

7. List the three categories of time delay rating for fuses and state a use for each type of time-delay rated fuse.

8. List the three categories of time delay rating for fuses and state a use for each type of time-delay rated fuse. Identify fuses as to voltage, current, and time delay ratings using fuses marked with the old military, new military, old commercial, and new commercial systems. List the three categories of time delay rating for fuses and state a use for each type of time-delay rated fuse.

9. Identify a clip-type and a post-type fuse holder from illustrations and identify the connections used on a post-type fuse holder for power source and load connections.

10. List the methods of checking for an open fuse, the items to check when replacing a fuse, the safety precautions to be observed when checking and replacing fuses, and the conditions to be checked for when conducting preventive maintenance on fuses.

11. Select a proper replacement and substitute fuse from a listing of fuses.

12. List the five main components of a circuit breaker and the three types of circuit breaker trip elements.

13. Describe the way in which each type of trip element reacts to excessive current.

14. Define the circuit breaker terms trip-free and nontrip-free and state one example for the use of each of these types of circuit breakers.

15. List the three time delay ratings of circuit breakers.

16. Define selective tripping, state why it is used, and state the way in which the time delay ratings of circuit breakers are used to design a selective tripping system.

17. Identify the factors used in selecting circuit breakers.

18. List the steps to follow before starting work on a circuit breaker and the items to be checked when maintaining circuit breakers.

CIRCUIT PROTECTION DEVICES

Electricity, like fire, can be either helpful or harmful to those who use it. A fire can keep people warm and comfortable when it is confined in a campfire or a furnace. It can be dangerous and destructive if it is on the loose and uncontrolled in the woods or in a building. Electricity can provide people with the light to read by or, in a blinding flash, destroy their eyesight. It can help save people's lives, or it can kill them. While we take advantage of the tremendous benefits electricity can provide, we must be careful to protect the people and systems that use it.

It is necessary then, that the mighty force of electricity be kept under control at all times. If for some reason it should get out of control, there must be a method of protecting people and equipment. Devices have been developed to protect people and electrical circuits from currents and voltages outside their normal operating ranges. Some examples of these devices are discussed in this chapter.

While you study this chapter, it should be kept in mind that a circuit protection device is used to keep an undesirably large current, voltage, or power surge out of a given part of an electrical circuit.

INTRODUCTION

An electrical unit is built with great care to ensure that each separate electrical circuit is fully insulated from all the others. This is done so that the current in a circuit will follow its intended path. Once the unit is placed into service, however, many things can happen to alter the original circuitry. Some of the changes can cause serious problems if they are not detected and corrected. While circuit protection devices cannot correct an abnormal current condition, they can indicate that an abnormal condition exists and protect personnel and circuits from that condition. In this chapter, you will learn what circuit conditions require protection devices and the types of protection devices used.

CIRCUIT CONDITIONS REQUIRING PROTECTION DEVICES

As has been mentioned, many things can happen to electrical and electronic circuits after they are in use. Chapter 1 of this module contains information showing you how to measure circuit characteristics to help determine the changes that can occur in them. Some of the changes in circuits can cause conditions that are dangerous to the circuit itself or to people living or working near the circuits. These potentially dangerous conditions require circuit protection. The conditions that require circuit protection are direct shorts, excessive current, and excessive heat.

Direct Short

One of the most serious troubles that can occur in a circuit is a DIRECT SHORT. Another term used to describe this condition is a SHORT CIRCUIT. The two terms mean the same thing and, in this chapter, the term direct short will be used. This term is used to describe a situation in which some point in the circuit, where full system voltage is present, comes in direct contact with the ground or return side of the circuit. This establishes a path for current flow that contains only the very small resistance present in the wires carrying the current.

According to Ohm's law, if the resistance in a circuit is extremely small, the current will be extremely large. Therefore, when a direct short occurs, there will be a very large current through the wires. Suppose, for instance, that the two leads from a battery to a motor came in contact with each other. If the leads were bare at the point of contact, there would be a direct short. The motor would stop running because all the current would be flowing through the short and none through the motor. The battery would become discharged quickly (perhaps ruined) and there could be the danger of fire or explosion.

The battery cables in our example would be large wires capable of carrying heavy currents. Most wires used in electrical circuits are smaller and their current carrying capacity is limited. The size of wire used in any given circuit is determined by space considerations, cost factors, and the amount of current the wire is expected to carry under normal operating conditions. Any current flow greatly in excess of normal, such as there would be in the case of a direct short, would cause a rapid generation of heat in the wire.

If the excessive current flow caused by the direct short is left unchecked, the heat in the wire will continue to increase until some portion of the circuit burns. Perhaps a portion of the wire will melt and open the circuit so that nothing is damaged other than the wire involved. The probability exists, however, that much greater damage will result. The heat in the wire can char and burn the insulation of the wire and that of other wires bundled with it, which can cause more shorts. If a fuel or oil leak is near any of the hot wires, a disastrous fire might be started.

Excessive Current

It is possible for the circuit current to increase without a direct short. If a resistor, capacitor, or inductor changes value, the total circuit impedance will also change in value. If a resistor _decreases_ in ohmic value, the total circuit resistance decreases. If a capacitor has a dielectric leakage, the capacitive reactance _decreases_. If an inductor has a partial short of its winding, inductive reactance decreases. Any of these conditions will cause an _increase_ in circuit current. Since the circuit wiring and components are designed to withstand normal circuit current, an increase in current would cause overheating (just as in the case of a direct short). Therefore, excessive current without a direct short will cause the same problems as a direct short.

Excessive Heat

As you have read, most of the problems associated with a direct short or excessive current concern the heat generated by the higher current. The damage to circuit components, the possibility of fire, and the possibility of hazardous fumes being given off from electrical components are consequences of excessive heat. It is possible for excessive heat to occur without a direct short or excessive current. If the bearings on a motor or generator were to fail, the motor or generator would overheat. If the temperature around an electrical or electronic circuit were to rise (through failure of a cooling system for example), excessive heat would be a problem. No matter what the cause, if excessive heat is present in a circuit, the possibility of damage, fire, and hazardous fumes exists.

> *Q1. Why are circuit protection devices necessary?*
>
> *Q2. What are the three conditions that require circuit protection?*
>
> *Q3. What is a direct short?*
>
> *Q4. What is an excessive current condition?*
>
> *Q5. What is an excessive heat condition?*

CIRCUIT PROTECTION DEVICES

All of the conditions mentioned are potentially dangerous and require the use of circuit protection devices. Circuit protection devices are used to stop current flow or open the circuit. To do this, a circuit protection device must ALWAYS be connected in series with the circuit it is protecting. If the protection device is connected in parallel, current will simply flow around the protection device and continue in the circuit.

A circuit protection device operates by opening and interrupting current to the circuit. The opening of a protection device shows that something is wrong in the circuit and should be corrected before the current is restored. When a problem exists and the protection device opens, the device should isolate the faulty circuit from the other unaffected circuits, and should respond in time to protect unaffected components in the faulty circuit. The protection device should NOT open during normal circuit operation.

The two types of circuit protection devices discussed in this chapter are fuses and circuit breakers.

Fuses

A fuse is the simplest circuit protection device. It derives its name from the Latin word "fusus," meaning "to melt." Fuses have been used almost from the beginning of the use of electricity. The earliest type of fuse was simply a bare wire between two connections. The wire was smaller than the conductor it was protecting and, therefore, would melt before the conductor it was protecting was harmed. Some "copper fuse link" types are still in use, but most fuses no longer use copper as the fuse element (the part of the fuse that melts). After changing from copper to other metals, tubes or enclosures were developed to hold the melting metal. The enclosed fuse made possible the addition of filler material, which helps to contain the arc that occurs when the element melts.

For many low power uses, the finer material is not required. A simple glass tube is used. The use of a glass tube gives the added advantage of being able to see when a fuse is open. Fuses of this type are commonly found in automobile lighting circuits.

Figure 2-1 shows several fuses and the symbols used on schematics.

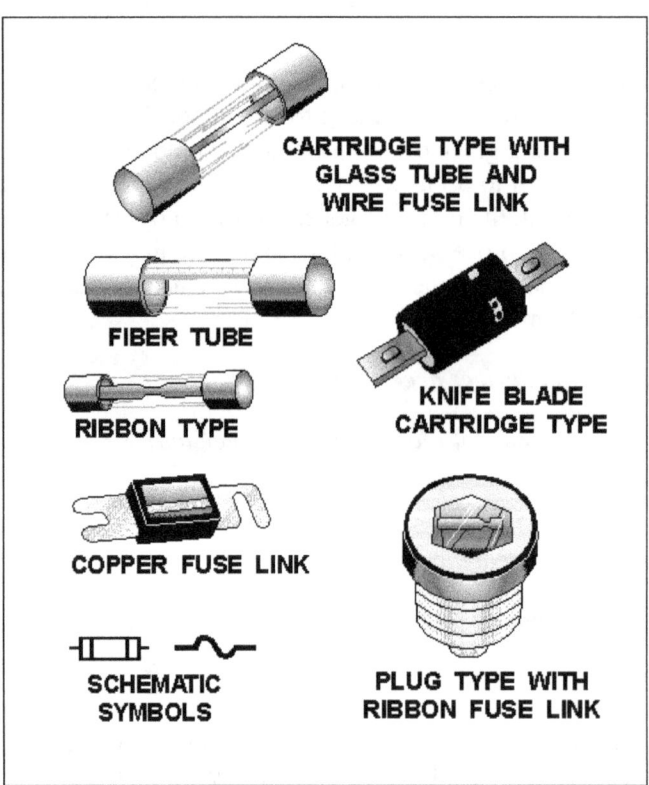

Figure 2-1.—Typical fuses and schematic symbols.

Circuit Breakers

While a fuse protects a circuit, it is destroyed in the process of opening the circuit. Once the problem that caused the increased current or heat is corrected, a new fuse must be placed in the circuit. A circuit protection device that can be used more than once solves the problems of replacement fuses. Such a device is safe, reliable, and tamper proof. It is also resettable, so it can be reused without replacing any parts. This device is called a CIRCUIT BREAKER because it breaks (opens) the circuit.

The first compact, workable circuit breaker was developed in 1923. It took 4 years to design a device that would interrupt circuits of 5000 amperes at 120 volts ac or dc. In 1928 the first circuit breaker was placed on the market. A typical circuit breaker and the appropriate schematic symbols are shown in figure 2-2.

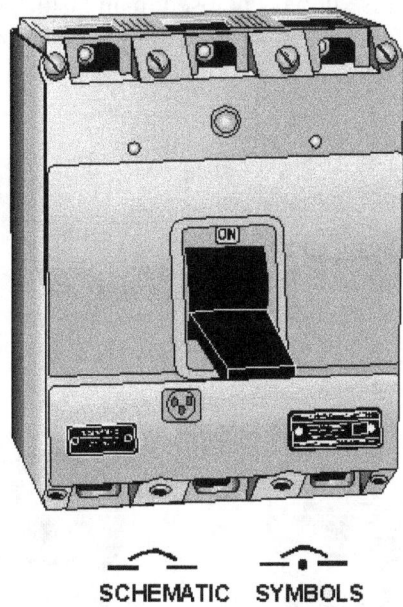

SCHEMATIC SYMBOLS

Figure 2-2.—Typical circuit breaker and schematic symbols.

Q6. *How are circuit protection devices connected to the circuit they are intended to protect and why are they connected in this way?*

Q7. *What are the two types of circuit protection devices?*

Q8. *Label the schematic symbols shown in figure 2-3 below.*

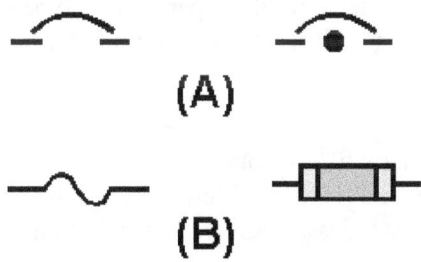

Figure 2-3.—Schematic symbols.

FUSE TYPES

Fuses are manufactured in many shapes and sizes. In addition to the copper fuse link already described, figure 2-1 shows other fuse types. While the variety of fuses may seem confusing, there are basically only two types of fuses: plug-type fuses and cartridge fuses. Both types of fuses use either a single wire or a ribbon as the fuse element (the part of the fuse that melts). The condition (good or bad) of some fuses can be determined by visual inspection. The condition of other fuses can only be determined with a meter. In the following discussion, visual inspection will be described. The use of meters to check fuses will be discussed later in this chapter.

PLUG-TYPE FUSE

The plug-type fuse is constructed so that it can be screwed into a socket mounted on a control panel or electrical distribution center. The fuse link is enclosed in an insulated housing of porcelain or glass. The construction is arranged so the fuse link is visible through a window of mica or glass. Figure 2-4 shows a typical plug-type fuse.

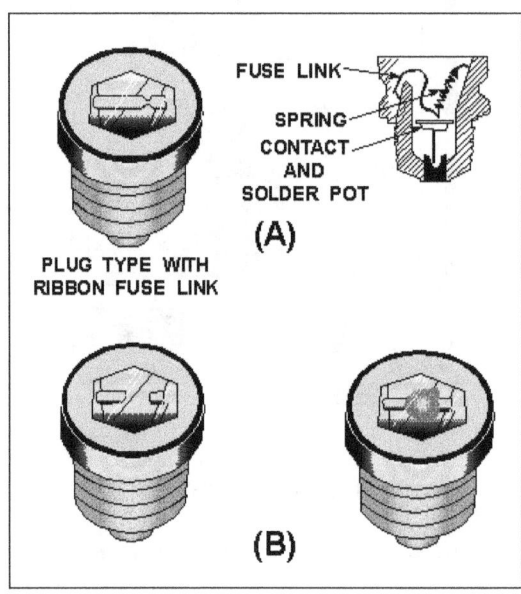

Figure 2-4.—Plug-type fuses:

Figure 2-4, view A, sows a good plug-type fuse. Notice the construction and the fuse link. In figure 2-4, view B, the same type of fuse is shown after the fuse link has melted. Notice the window showing the indication of this open fuse. The indication could be either of the ones shown in figure 2-4, view B.

The plug-type fuse is used primarily in low-voltage, low-current circuits. The operating range is usually up to 150 volts and from 0.5 ampere to 30 amperes. This type of fuse is found in older circuit protection devices and is rapidly being replaced by the circuit breaker.

CARTRIDGE FUSE

The cartridge fuse operates exactly like the plug-type fuse. In the cartridge fuse, the fuse link is enclosed in a tube of insulating material with metal ferrules at each end (for contact with the fuse holder). Some common insulating materials are glass, bakelite, or a fiber tube filled with insulating powder.

Figure 2-5 shows a glass-tube fuse. In figure 2-5, view A, notice the fuse link and the metal ferrules. Figure 2-5, view B, shows a glass-tube fuse that is open. The open fuse link could appear either of the ways shown in figure 2-5, view B.

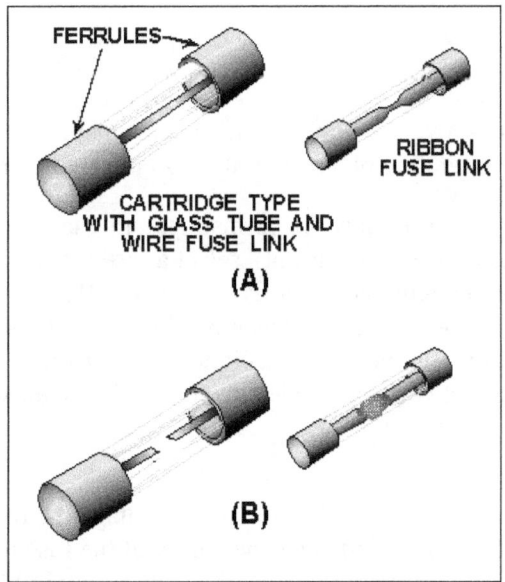

Figure 2-5.—Cartridge-tube fuse.

Cartridge fuses are available in a variety of physical sizes and are used in many different circuit applications. They can be rated at voltages up to 10,000 volts and have current ratings of from 1/500 (.002) ampere to 800 amperes. Cartridge fuses may also be used to protect against excessive heat and open at temperatures of from 165° F to 410°F (74°C to 210°C).

Q9. Label the fuses shown in figure 2-6 according to type.

Q10. Identify the open fuses shown in figure 2-6.

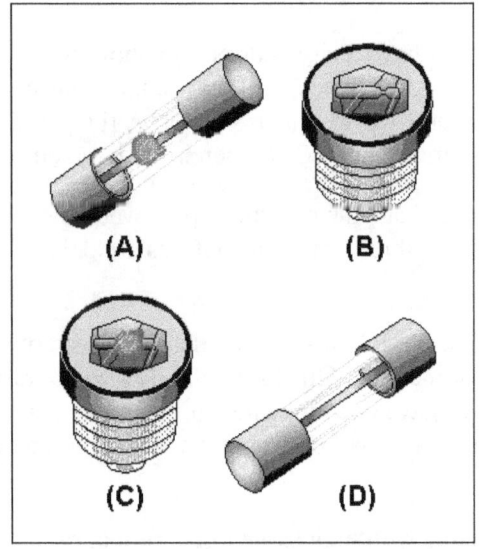

Figure 2-6.—Fuse recognition.

FUSE RATINGS

You can determine the physical size and type of a fuse by looking at it, but you must know other things about a fuse to use it properly. Fuses are rated by current, voltage, and time-delay characteristics to aid in the proper use of the fuse. To select the proper fuse, you must understand the meaning of each of the fuse ratings.

CURRENT RATING

The current rating of a fuse is a value expressed in amperes that represents the current the fuse will allow without opening. The current rating of a fuse is always indicated on the fuse.

To select the proper fuse, you must know the normal operating current of the circuit. If you wish to protect the circuit from overloads (excessive current), select a fuse rated at 125 percent of the normal circuit current. In other words, if a circuit has a normal current of 10 amperes, a 12.5-ampere fuse will provide overload protection. If you wish to protect against direct shorts only, select a fuse rated at 150 percent of the normal circuit current. In the case of a circuit with 10 amperes of current, a 15 ampere fuse will protect against direct shorts, but will not be adequate protection against excessive current.

VOLTAGE RATING

The voltage rating of a fuse is NOT an indication of the voltage the fuse is designed to withstand while carrying current. The voltage rating indicates the ability of the fuse to quickly extinguish the arc after the fuse element melts and the maximum voltage the open fuse will block. In other words, once the fuse has opened, any voltage less than the voltage rating of the fuse will not be able to "jump" the gap of the fuse. Because of the way the voltage rating is used, it is a maximum rms voltage value. You must always select a fuse with a voltage rating equal to or higher than the voltage in the circuit you wish to protect.

TIME DELAY RATING

There are many kinds of electrical and electronic circuits that require protection. In some of these circuits, it is important to protect against temporary or transient current increases. Sometimes the device being protected is very sensitive to current and cannot withstand an increase in current. In these cases, a fuse must open very quickly if the current increases.

Some other circuits and devices have a large current for short periods and a normal (smaller) current most of the time. An electric motor, for instance, will draw a large current when the motor starts, but normal operating current for the motor will be much smaller. A fuse used to protect a motor would have to allow for this large temporary current, but would open if the large current were to continue.

Fuses are time delay rated to indicate the relationship between the current through the fuse and the time it takes for the fuse to open. The three time delay ratings are delay, standard, and fast.

Delay

A delay, or slow-blowing, fuse has a built-in delay that is activated when the current through the fuse is greater than the current rating of the fuse. This fuse will allow temporary increases in current (surge) without opening. Some delay fuses have two elements; this allows a very long time delay. If the overcurrent condition continues, a delay fuse will open, but it will take longer to open than a standard or a fast fuse.

Delay fuses are used for circuits with high surge or starting currents, such as motors, solenoids, and transformers.

Standard

Standard fuses have no built-in time delay. Also, they are not designed to be very fast acting. Standard fuses are sometimes used to protect against direct shorts only. They may be wired in series with a delay fuse to provide faster direct short protection. For example, in a circuit with a 1-ampere delay fuse, a 5-ampere standard fuse may be used in addition to the delay fuse to provide faster protection against a direct short.

A standard fuse can be used in any circuit where surge currents are not expected and a very fast opening of the fuse is not needed. A standard fuse opens faster than a delay fuse, but slower than a fast rated fuse.

Standard fuses can be used for automobiles, lighting circuits, or electrical power circuits.

Fast

Fast fuses are designed to open very quickly when the current through the fuse exceeds the current rating of the fuse. Fast fuses are used to protect devices that are very sensitive to increased current. A fast fuse will open faster than a delay or standard fuse.

Fast fuses can be used to protect delicate instruments or semiconductor devices.

Figure 2-7 will help you understand the differences between delay, standard, and fast fuses. Figure 2-7 shows that, if a 1-ampere rated fuse had 2 amperes of current through it, (200% of the rated value), a fast fuse would open in about .7 second, a standard rated fuse would open in about 1.5 seconds, and a delay rated fuse would open in about 10 seconds. Notice that in each of the fuses, the time required to open the fuse decreases as the rated current increases.

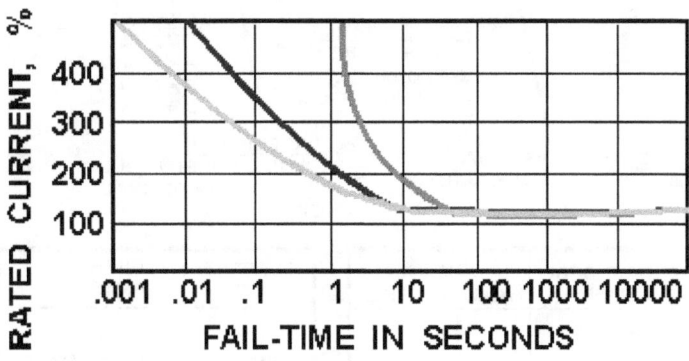

Figure 2-7.—Time required for fuse to open.

Q11. In what three ways are fuses rated?

Q12. What does the current rating of a fuse indicate?

Q13. What does the voltage rating of a fuse indicate?

Q14. What are the three time delay ratings of fuses?

Q15. Give an example of a device you could protect with each type of time delay fuse.

IDENTIFICATION OF FUSES

Fuses have identifications printed on them. The printing on the fuse will identify the physical size, the type of fuse, and the fuse ratings. There are four different systems used to identify fuses. The systems are the old military designation, the new military designation, the old commercial designation, and the new commercial designation. All four systems are presented here, so you will be able to identify a fuse no matter which designation is printed on the fuse.

You may have to replace an open fuse that is identified by one system with a good fuse that is identified by another system. The designation systems are fairly simple to understand and cross-reference once you are familiar with them.

OLD MILITARY DESIGNATION

Figure 2-8 shows a fuse with the old military designation. The tables in the lower part of the figure show the voltage and current codes used in this system. The upper portion of the figure is the explanation of the old military designation. The numbers and letters in parentheses are the coding for the fuse shown in figure 2-8.

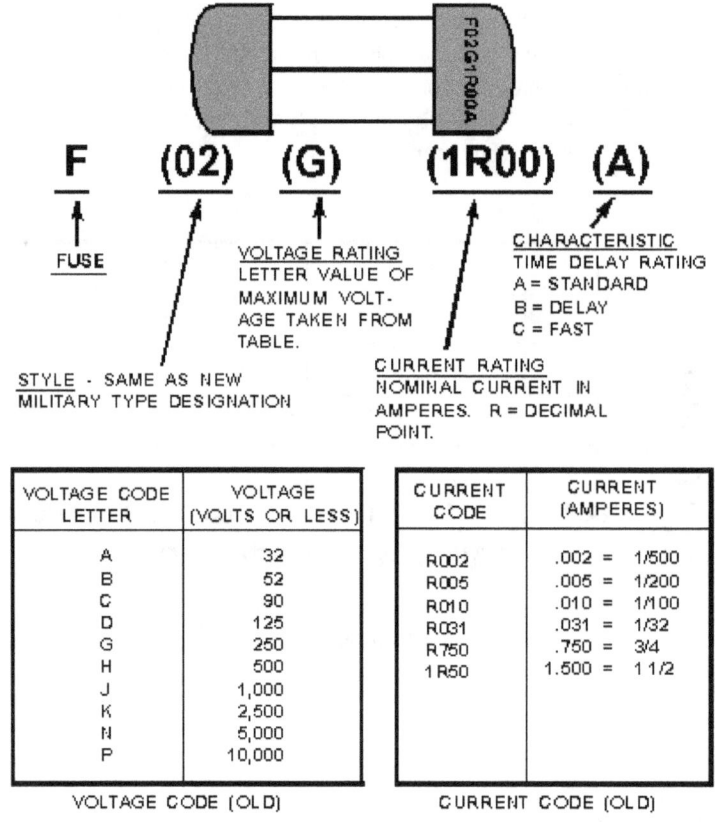

Figure 2-8.—Old type military fuse designation.

The old military designation always starts with "F," which stands for fuse. Next, the set of numbers (02) indicates the style. Style means the construction and dimensions (size) of the fuse. Following the style is a letter that represents the voltage rating of the fuse (G). The voltage code table in figure 2-8 shows each voltage rating letter and its meaning in volts. In the example shown, the voltage ratings is G, which means the fuse should be used in a circuit where the voltage is 250 volts or less. After this is a set of three numbers and the letter "R," which represent the current rating of the fuse. The "R" indicates the decimal point. In the example shown, the current rating is 1R00 or 1.00 ampere. Some other examples of the current rating are shown in the current code table of figure 2-8. The final letter in the old military designation (A) indicates the time delay rating of the fuse.

While the old military designation is still found on some fuses, the voltage and current ratings must be "translated," since they use letters to represent numerical values. The military developed the new military designations to make fuse identification easier.

NEW MILITARY DESIGNATION

Figure 2-9 is an example of a fuse coded in the new military designation. The fuse identified in the example in figure 2-9 is the same type as the fuse used as an example in figure 2-8.

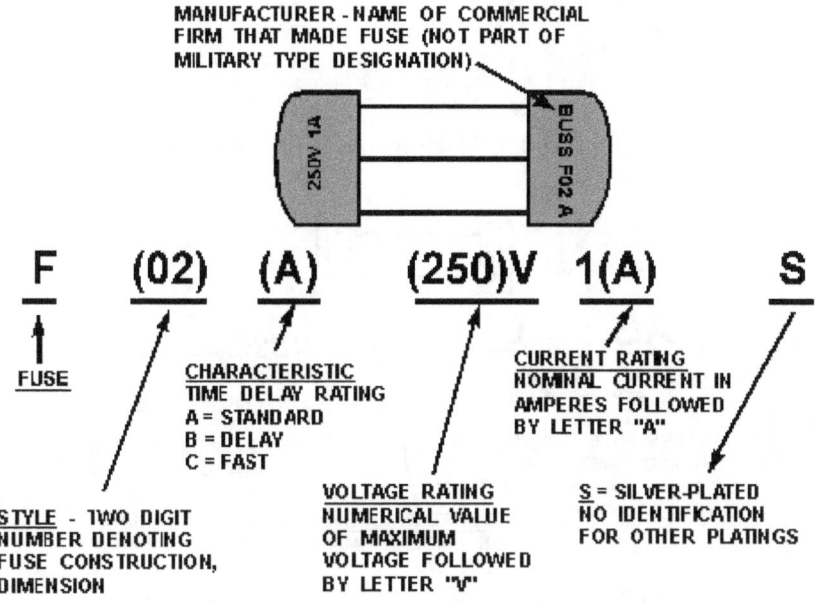

Figure 2-9.—New type military fuse designation.

The new military designation always start with the letter "F," which stands for fuse. The set of numbers (02) next to this indicates the style. The style numbers are identical to the ones used in the old military designation and indicate the construction and dimensions of the fuse. Following the style designation is a single letter (A) that indicates the time delay rating of the fuse. This is the same time delay rating code as indicated in the old military designation, but the position of this letter in the coding is changed to avoid confusing the "A" for standard time delay with the "A" for ampere. Following the time delay rating is the voltage rating of the fuse (250) V. In the old military designation, a letter was used to indicate the voltage rating. In the new military designation, the voltage is indicated by numbers followed by a "V," which stands for volts or less. After the voltage rating, the current rating is given by numbers followed by the letter "A." The current rating may be a whole number (1A), a fraction (1/500 A), a whole number and a fraction (1 1/2A), a decimal (0.250A), or a whole number and a decimal (1.50A). If the ferrules of the fuse are silver-plated, the current rating will be followed by the letter "S." If any other plating is used, the current rating will be the last part of the fuse identification.

As you can see, the new military designation is much easier to understand than the old military designation.

You may find a fuse coded in one of the commercial designations. The commercial designations are fairly easy to understand and figure 2-10 shows the old and new commercial designations for the same type of fuse that was used in figures 2-8 and 2-9.

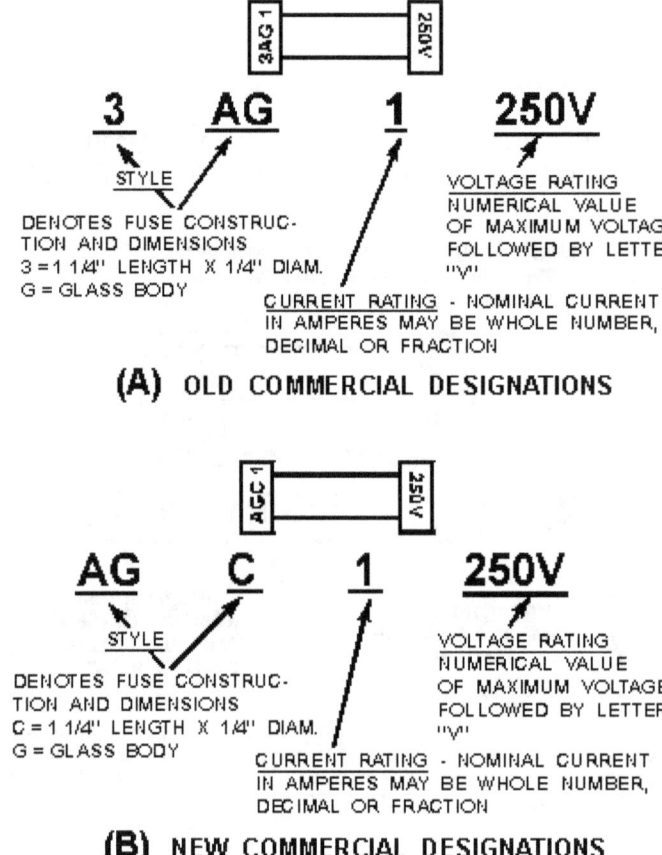

Figure 2-10.—Commercial designations for fuses:

OLD COMMERCIAL DESIGNATION

Figure 2-10, view A, shows the old commercial designation for a fuse. The first part of the designation is a combination of letters and numbers (three in all) that indicates the style and time delay characteristics. This part of the designation (3AG) is the information contained in the style and time delay rating portions of military designations.

In the example shown, the code 3AG represents the same information as the underlined portions of F02 G 1R00 A from figure 2-8 (Old Military Designation) and F02A 250VIAS from figure 2-9 (New Military Designation). The only way to know the time delay rating of this fuse is to look it up in the manufacturer's catalog or in a cross-reference listing to find the military designation. The catalog will tell you the physical size, the material from which the fuse is constructed, and the time delay rating of the fuse. A 3AG fuse is a glass-bodied fuse, 1/4 inch × 1 1/4 inches (6.35 millimeters × 31.8 millimeters) and has a standard time delay rating.

Following the style designation is a number that is the current rating of the fuse (1). This could be a whole number, a fraction, a whole number and a fraction, a decimal, or a whole number and a decimal. Following the current rating is the voltage rating; which, in turn, is followed by the letter "V," which stands for volts or less (250V).

NEW COMMERCIAL DESIGNATION

Figure 2-10, view B, shows the new commercial designation for fuses. It is the same as the old commercial designation except for the style portion of the coding. In the old commercial system, the style was a combination of letters and numbers. In the new commercial system, only letters are used. In the example shown, 3AG in the old system becomes AGC in the new system. Since "C" is the third letter of the alphabet, it is used instead of the "3" used in the old system. Once again, the only way to find out the time delay rating is to look up this coding in the manufacturer's catalog or to use a cross-reference listing. The remainder of the new commercial designation is exactly the same as the old commercial designation.

Q16. What are the voltage, current, and time delay ratings for a fuse with the designation

(a) F02DIR50B?

(b) $F02A250V\frac{1}{8}A$?

Q17. What are the voltage and current ratings for a fuse designated

(a) $3AG\frac{1}{16}125V$?

(b) $FNA\frac{15}{100}250V$?

Q18. What is the new military designation for a fuse with the old military designation F05A20ROB?

FUSEHOLDERS

For a fuse to be useful, it must be connected to the circuit it will protect. Some fuses are "wired in" or soldered to the wiring of circuits, but most circuits make use of FUSEHOLDERS. A fuseholder is a device that is wired into the circuit and allows easy replacement of the fuse.

Fuseholders are made in many shapes and sizes, but most fuseholders are basically either clip-type or post-type. Figure 2-11 shows a typical clip-type and post-type fuseholder.

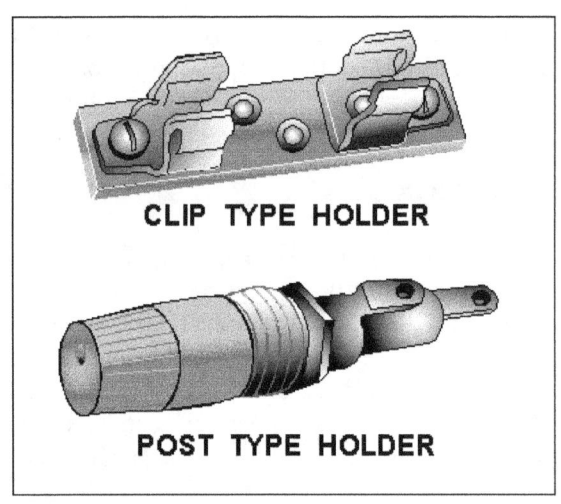

Figure 2-11.—Typical fuseholders.

CLIP-TYPE FUSEHOLDER

The clip-type fuseholder is used for cartridge fuses. The ferrules or knife blade of the fuse are held by the spring tension of the clips. These clips provide the electrical connection between the fuse and the circuit. If a glass-bodied fuse is used, the fuse can be inspected visually for an open without removing the fuse from the fuse holder. Clip-type fuseholders are made in several sizes to hold the many styles of fuses. The clips maybe made for ferrules or knife blade cartridge fuses. While the base of a clip-type fuseholder is made from insulating material, the clips themselves are conductors. The current through the fuse goes through the clips and care must be taken to not touch the clips when there is power applied. If the clips are touched, with power applied, a severe shock or a short circuit will occur.

POST-TYPE FUSEHOLDERS

Post-type fuseholders are made for cartridge fuses. The post-type fuseholder is much safer because the fuse and fuse connections are covered with insulating material. The disadvantage of the post-type fuseholder is that the fuse must be removed to visually check for an open. The post-type fuseholder has a cap that screws onto the body of the fuseholder. The fuse is held in this cap by a spring-type connector and, as the cap is screwed on, the fuse makes contact with the body of the fuseholder. When the cap and fuse are removed from the body of the fuseholder, the fuse is removed from the circuit and there is no danger of shock or short circuit from touching the fuse.

Post-type fuseholders are usually mounted on the chassis of the equipment in which they are used. After wires are connected to the fuseholder, insulating sleeves are placed over the connections to reduce the possibility of a short circuit. Notice the two connections on the post-type fuseholder of figure 2-11. The connection on the right is called the center connector. The other connector is the outside connector. The outside connector will be closer to the equipment chassis. (The threads and nut shown are used to fasten the fuseholder to the chassis.) The possibility of the outside connector coming in contact with the chassis (causing a short circuit) is much higher than the possibility of the center conductor contacting the chassis. The power source should always be connected to the center connector so the fuse will open if the outside connector contacts the chassis. If the power source were connected to the outside connector, and the outside connector contacted the chassis, there would be a direct short, but the fuse would not open.

Q19. Label the fuseholders in figure 2-12.

Q20. Which connector should you use to connect the (a) power source and (b) load to the fuseholder shown in figure 2-12(A)?

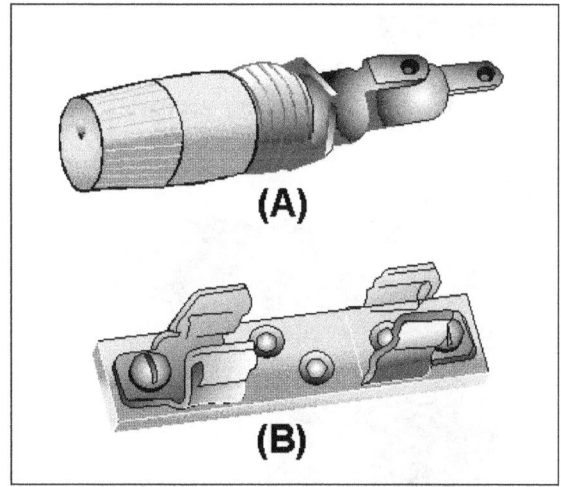

Figure 2-12.—Fuseholder identification.

CHECKING AND REPLACEMENT OF FUSES

A fuse, if properly used, should not open unless something is wrong in the circuit the fuse is protecting. When a fuse is found to be open, you must determine the reason the fuse is open. Replacing the fuse is not enough.

Before you look for the cause of an open fuse, you must be able to determine if the fuse is open.

CHECKING FOR AN OPEN FUSE

There are several ways of checking for an open fuse. Some fuses and fuseholders have indicators built in to help you find an open fuse; also, a multimeter can be used to check fuses. The simplest way to check glass-bodied fuses, and the method you should use first, is visual inspection.

Visual Inspection

An open glass-bodied fuse can usually be found by visual inspection. Earlier in this chapter, figures 2-4 and 2-5 showed you how an open plug-type and an open glass-bodied cartridge-type fuse would look. If the fuse element is not complete, or if the element has been melted onto the glass tube, the fuse is open.

It is not always possible to tell if a fuse is open by visual inspection. Fuses with low current ratings have elements that are so small, it is sometimes not possible to know if the fuse link is complete simply by looking at it. If the fuse is not glass-bodied, it will not be possible to check the fuse visually. Also, sometimes a fuse will look good, but will, in fact, be open. Therefore, while it is sometimes possible to know if a fuse is open by visual inspection, it is not possible to be sure a fuse is good just by looking at it.

Fuse Indicators

Some fuses and fuseholders have built-in indicators to show when a fuse is open. Examples of these open-fuse indicators are shown in figure 2-13. Figure 2-13, view A, shows a cartridge-type fuse with an open-fuse indicator. The indicator is spring loaded and held by the fuse link. If the fuse link opens, the spring forces the indicator out. Some manufacturers color the indicator so it is easier to see in the open-fuse position.

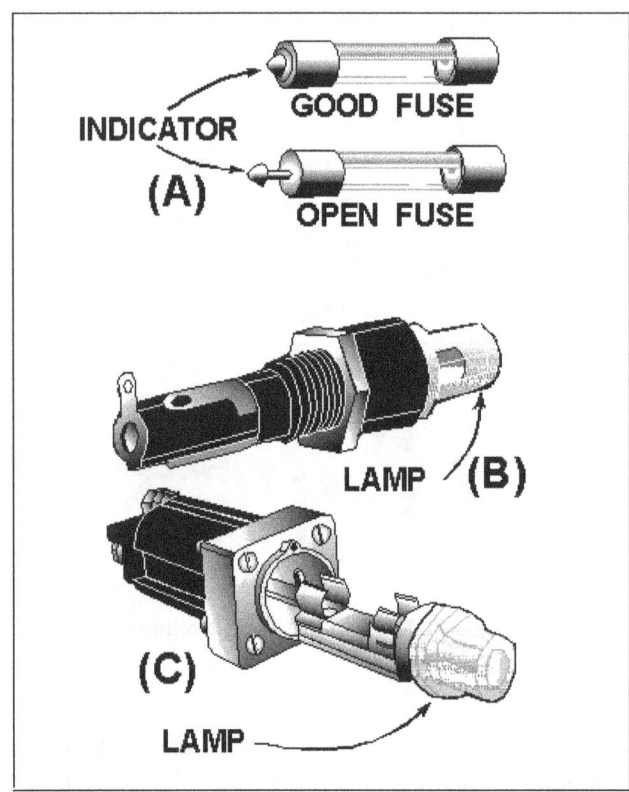

Figure 2-13.—Open fuse indicators: Clip-type fuseholder with an indicating lamp.

Figure 2-13, view B, shows a plug-type fuseholder with an indicating lamp in the fuse cap. If the fuse opens, the lamp in the fuse cap will light. Figure 2-13, view C, shows a clip-type fuseholder with an indicating lamp.

Just as in visual checking, the indicator can show an open fuse. Since the indicator may not always work, you cannot be sure a fuse is good just because there is no open-fuse indication.

Checking Fuses with a Meter

The only sure method of determining if a fuse is open is to use a meter. An ohmmeter can be used to check for an open fuse by removing the fuse from the circuit and checking for continuity through the fuse (0 ohms). If the fuse is not removed from the circuit, and the fuse is open, the ohmmeter may measure the circuit resistance. This resistance reading might lead you to think the fuse is good. You must be careful when you use an ohmmeter to check fuses with small current ratings (such as 1/32 ampere or less), because the current from the ohmmeter may be larger than the current rating of the fuse. For most practical uses, a small current capacity fuse can be checked out of the circuit through the use of a resistor. The ohmic value of the resistor is first measured and then placed in series with the fuse. The continuity reading on the ohmmeter should be of the same value, or close to it, as the original value of the resistor. This method provides protection for the fuse by dropping the voltage across the resistor. This in turn decreases the power in the form of heat at the fuse. Remember, it is heat which melts the fuse element.

A voltmeter can also be used to check for an open fuse. The measurement is taken between each end of the fuse and the common or ground side of the line. If voltage is present on <u>both</u> sides of the fuse (from the voltage source and to the load), the fuse is not open. Another method commonly used, is to measure <u>across</u> the fuse with the voltmeter. If NO voltage is indicated on the meter, the fuse is good, (not open).

Remember there is no voltage drop across a straight piece of wire. Some plug-type fuseholders have test points built in to allow you to check the voltage. To check for voltage on a clip-type fuseholder, check each of the clips. The advantage of using a voltmeter to check for an open fuse is that the circuit does not have to be deenergized and the fuse does not have to be removed.

WARNING

PERSONNEL MAY BE EXPOSED TO HAZARDOUS VOLTAGE

Safety Precautions When Checking a Fuse

Since a fuse has current through it, you must be very careful when checking for an open fuse to avoid being shocked or damaging the circuit. The following safety precautions will protect you and the equipment you are using.

- Turn the power off and discharge the circuit before removing a fuse.

- Use a fusepuller (an insulated tool) when you remove a fuse from a clip-type fuseholder.

- When you check a fuse with a voltmeter, be careful to avoid shocks and short circuits.

- When you use an ohmmeter to check fuses with low current ratings, be careful to avoid opening the fuse by excessive current from the ohmmeter.

Q21. What are three methods for determining if a fuse is open?

Q22. You have just checked a fuse with an ohmmeter and find that the fuse is shorted. What should you do?

Q23. You have just checked a 1/500-ampere fuse with an ohmmeter and find it is open. Checking the replacement fuse shows the replacement fuse is open also. Why would the replacement fuse indicate open?

Q24. How could you check a 1/500-ampere fuse with an ohmmeter?

Q25. List the safety precautions to be observed when checking fuses.

REPLACEMENT OF FUSES

After an open fuse is found and the trouble that caused the fuse to open has been corrected, the fuse must be replaced. Before you replace the fuse, you must be certain the replacement fuse is the proper type and fits correctly.

Proper Type of Replacement Fuse

To be certain a fuse is the proper type, check the technical manual for the equipment. The parts list will show you the proper fuse identification for a replacement fuse. Obtain the exact fuse specified, if possible, and check the identification number of the replacement fuse against the parts list.

If you cannot obtain a direct replacement, use the following guidelines:

- Never use a fuse with a higher current rating, a lower voltage rating, or a slower time delay rating than the specified fuse.

- The best substitution fuse is a fuse with the same current and time delay ratings and a higher voltage rating.

- If a lower current rating or a faster time delay rating is used, the fuse may open under normal circuit conditions.

- Substitute fuses must have the same style (physical dimensions) as the specified fuse.

Proper Fit of Replacement Fuses

When you have obtained a proper replacement fuse, you must make certain it will fit correctly in the fuseholder. If the fuseholder is corroded, the fuse will not fit properly. In addition, the corrosion can cause increased resistance or heating. Clean corroded terminals with fine sandpaper so that all corrosion is removed. Do **NOT** lubricate the terminals. If the terminals are badly pitted, replace the fuseholder. Be certain the replacement fuseholder is the correct size and type by checking the parts list in the technical manual for the equipment.

After you check for and correct any corrosion problems, be certain the fuse fits tightly in the fuseholder. When you insert the fuse in the cap of a plug-type fuseholder, the fuse should fit tightly. A small amount of pressure should be needed to insert the fuse and cap into the fuseholder body.

In clip-type fuseholders, the clips can be easily bent out of shape. This causes an incorrect fit, which in time could cause an equipment malfunction. Figure 2-14 shows examples of correct and incorrect fuse contacts for clip-type fuseholders used with knifeblade and ferrule cartridge fuses. The clips shown in the left picture of each row have the correct contact. The three pictures on the right of each row show incorrect contact. Notice how the clips are not contacting completely with the knifeblade or ferrules. This incomplete contact can. cause corrosion at the contacts, which in turn can create a high resistance and drop some of the circuit voltage at this point.

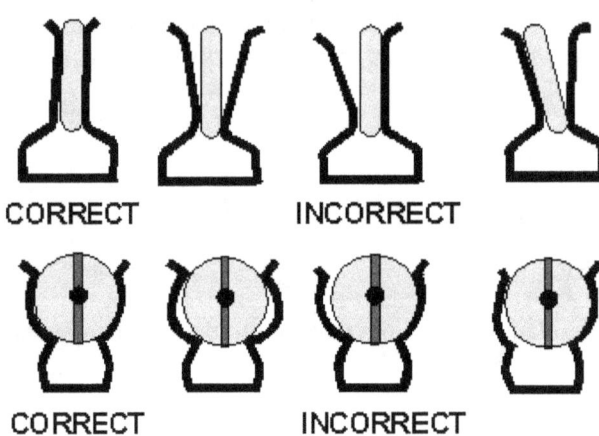

Figure 2-14.—Contact between clips and fuses.

If the fuse clips do not make complete contact with the fuse, try to bend the clips back into shape. If the clips cannot be repaired by bending, replace the fuseholder or use clip clamps. Clip clamps are shown in figure 2-15.

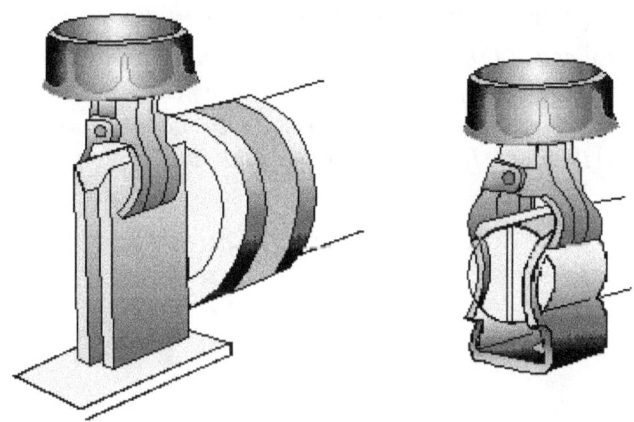

Figure 2-15.—Clip clamps.

Safety Precautions When Replacing Fuses

The following safety precautions will prevent injury to personnel and damage to equipment. These are the MINIMUM safety precautions for replacing fuses.

- Be sure the power is off in the circuit and the circuit is discharged before replacing a fuse.

- Use an identical replacement fuse if possible.

- Remove any corrosion from the fuseholder before replacing the fuse.

- Be certain the fuse properly fits the fuseholder.

PREVENTIVE MAINTENANCE OF FUSES

Preventive maintenance of fuses consists of checking for the following conditions and correcting any discrepancies.

1. IMPROPER FUSE. Check the fuse installed against that recommended in the technical manual for the equipment. If an incorrect fuse is installed, replace it with the correct fuse.

2. CORROSION. Check for corrosion on the fuseholder terminals or the fuse itself. If corrosion is present, remove it with fine sandpaper.

3. IMPROPER FIT. Check for contact between the fuse and fuseholder. If a piece of paper will fit between the fuse and the clips on a clip-type fuseholder, there is improper contact. If the fuse is not held in the cap of a plug-type fuseholder, the contacts are too loose.

4. OPEN FUSES. Check fuses for opens. If any fuse is open, repair the trouble that caused the open fuse and replace the fuse.

Q26. You have removed an open fuse from a fuseholder and repaired the cause of the fuse opening. The parts list specifies a fuse coded F02BI25V½A. There are no fuses available with that identification. In the following list, indicate if the fuse is a direct replacement, a good substitute, or not acceptable. For the fuses that are good substitutes, number them in order of preference and explain why they are numbered that way. If the fuse is not acceptable, explain why.

(a) F03BI25V½A

(b) F02BI25V⅜A

(c) F02GR500B

(d) F02B32V½A

(e) F02DR500B

(f) F02A250V⅝A

(g) F02AI25V½A

Q27. What two things should you check before replacing a fuse?

Q28. List the safety precautions to be observed when replacing a fuse.

Q29. What conditions should you check for when conducting preventive maintenance on fuses?

CIRCUIT BREAKERS

A circuit breaker is a circuit protection device that, like a fuse, will stop current in the circuit if there is a direct short, excessive current, or excessive heat. Unlike a fuse, a circuit breaker is reusable. The circuit breaker does not have to be replaced after it has opened or broken the circuit. Instead of replacing the circuit breaker, you reset it.

Circuit breakers can also be used as circuit control devices. By manually opening and closing the contacts of a circuit breaker, you can switch the power on and off. Circuit control devices will be covered in more detail in the next chapter.

Circuit breakers are available in a great variety of sizes and types. It would not be possible to describe every type of circuit breaker in use today, but this chapter will describe the basic types of circuit breakers and their operational principles.

Circuit breakers have five main components, as shown in figure 2-16. The components are the frame, the operating mechanism, the arc extinguishers and contacts, the terminal connectors, and the trip elements.

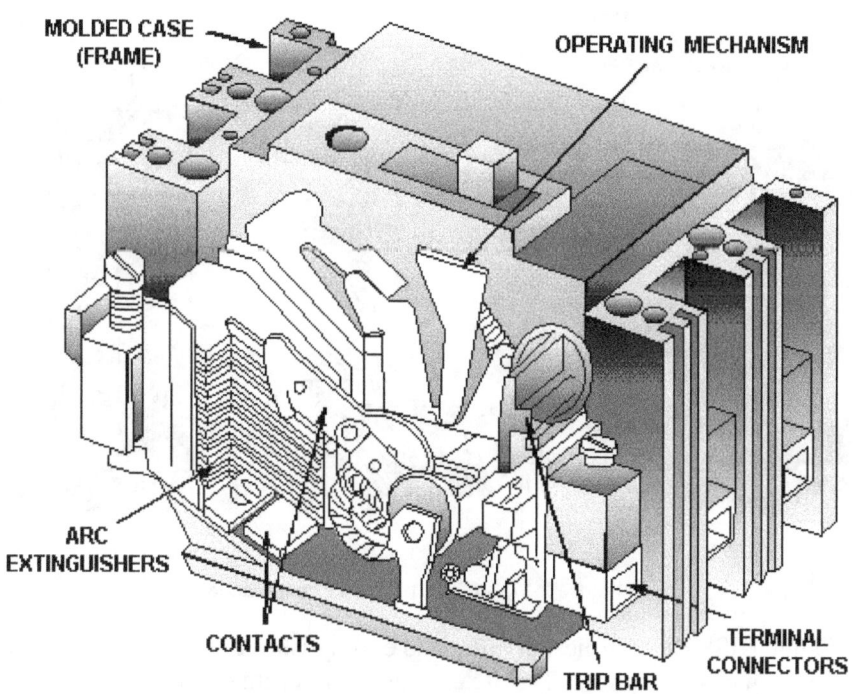

Figure 2-16.—Circuit breaker components.

The FRAME provides an insulated housing and is used to mount the circuit breaker components (fig. 2-17). The frame determines the physical size of the circuit breaker and the maximum allowable voltage and current.

The OPERATING MECHANISM provides a means of opening and closing the breaker contacts (turning, the circuit ON and OFF). The toggle mechanism shown in figure 2-17 is the quick-make, quick-break type, which means the contacts snap open or closed quickly, regardless of how fast the handle is moved. In addition to indicating whether the breaker is ON or OFF, the operating mechanism handle indicates when the breaker has opened automatically (tripped) by moving to a position between ON and OFF. To reset the circuit breaker, the handle must first be moved to the OFF position, and then to the ON position.

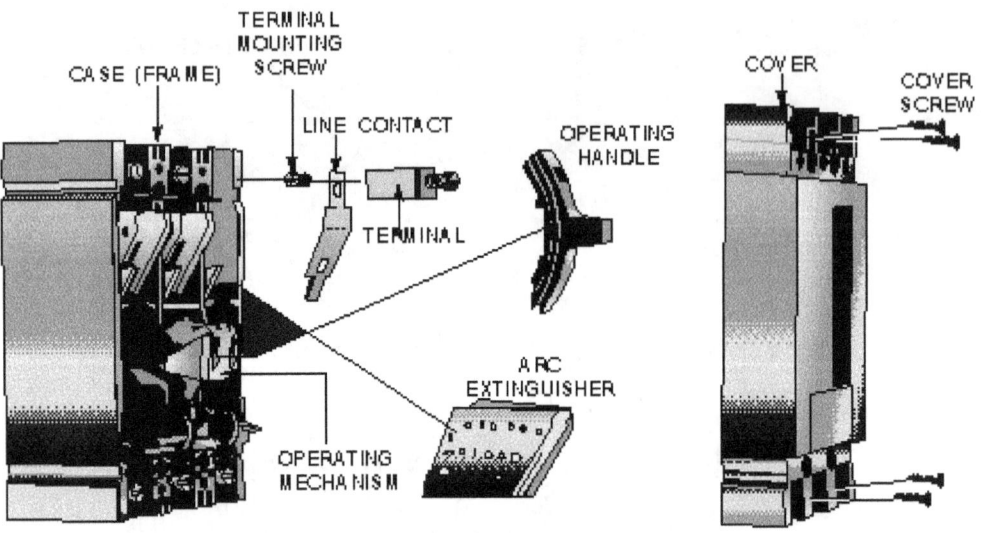

Figure 2-17.—Circuit breaker construction.

The ARC EXTINGUISHER confines, divides, and extinguishes the arc drawn between contacts each time the circuit breaker interrupts current. The arc extinguisher is actually a series of contacts that open gradually, dividing the arc and making it easier to confine and extinguish. This is shown in figure 2-18. Arc extinguishers are generally used in circuit breakers that control a large amount of power, such as those found in power distribution panels. Small power circuit breakers (such as those found in lighting panels) may not have arc extinguishers.

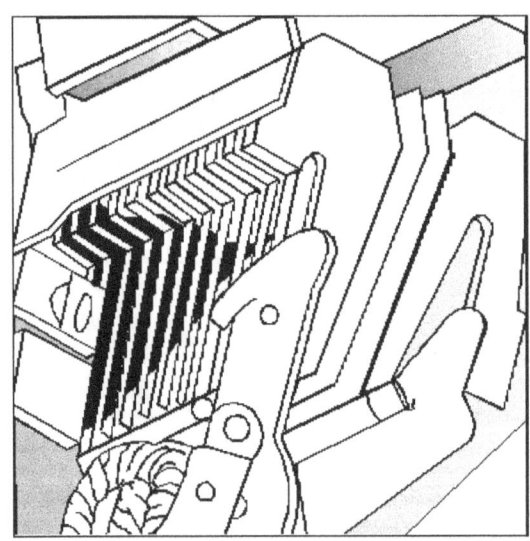

Figure 2-18.—Arc extinguisher action.

TERMINAL CONNECTORS are used to connect the circuit breaker to the power source and the load. They are electrically connected to the contacts of the circuit breaker and provide the means of connecting the circuit breaker to the circuit.

The TRIP ELEMENT is the part of the circuit breaker that senses the overload condition and causes the circuit breaker to trip or break the circuit. This chapter will cover the thermal, magnetic, and thermal-magnetic trip units used by most circuit breakers. (Some circuit breakers make use of solid-state trip units using current transformers and solid-state circuitry.)

THERMAL TRIP ELEMENT

A thermal trip element circuit breaker uses a bimetallic element that is heated by the load current. The bimetallic element is made from strips of two different metals bonded together. The metals expand at different rates as they are heated. This causes the bimetallic element to bend as it is heated by the current going to the load. Figure 2-19 shows how this can be used to trip the circuit breaker.

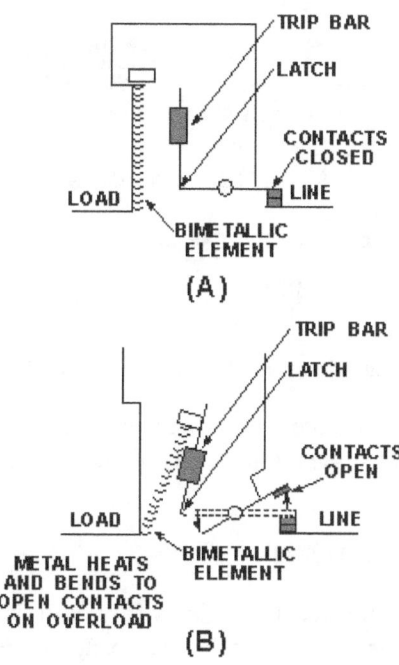

Figure 2-19.—Thermal trip element action: A. Trip element with normal current; B. Contacts open.

Figure 2-19, view A, shows the trip element with normal current. The bimetallic element is not heated excessively and does not bend. If the current increases (or the temperature around the circuit breaker increases), the bimetallic element bends, pushes against the trip bar, and releases the latch. Then, the contacts open, as shown in figure 2-19, view B.

The amount of time it takes for the bimetallic element to bend and trip the circuit breaker depends on the amount the element is heated. A large overload will heat the element quickly. A small overload will require a longer time to trip the circuit breaker.

MAGNETIC TRIP ELEMENT

A magnetic trip element circuit breaker uses an electromagnet in series with the circuit load as in figure 2-20. With normal current, the electromagnet will not have enough attraction to the trip bar to move it, and the contacts will remain closed as shown in figure 2-20, view A. The strength of the magnetic field of the electromagnet increases as current through the coil increases. As soon as the current in the circuit becomes large enough, the trip bar is pulled toward the magnetic element (electromagnet), the contacts are opened, and the current stops, as shown in figure 2-20, view B.

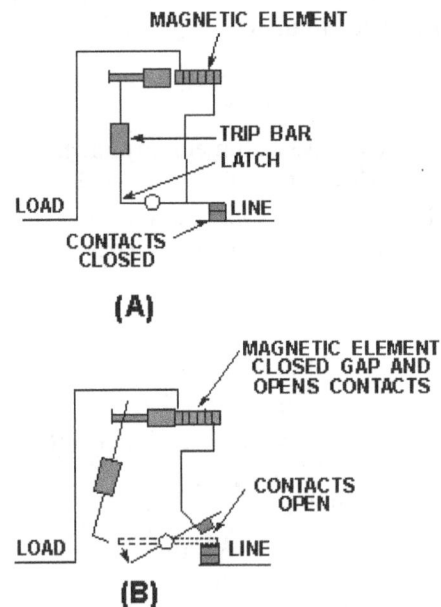

Figure 2-20.—Magnetic trip element action; Closed contacts;

The amount of current needed to trip the circuit breaker depends on the size of the gap between the trip bar and the magnetic element. On some circuit breakers, this gap (and therefore the trip current) is adjustable.

THERMAL-MAGNETIC TRIP ELEMENT

The thermal trip element circuit breaker, like a delay fuse, will protect a circuit against a small overload that continues for a long time. The larger the overload, the faster the circuit breaker will trip. The thermal element will also protect the circuit against temperature increases. A magnetic circuit breaker will trip instantly when the preset current is present. In some applications, both types of protection are desired. Rather than use two separate circuit breakers, a single trip element combining thermal and magnetic trip elements is used. A thermal-magnetic trip element is shown in figure 2-21.

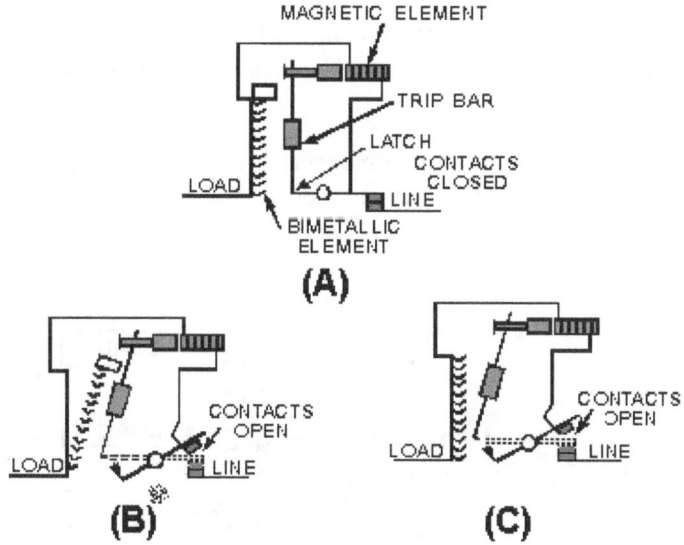

Figure 2-21.—Thermal-magnetic element action:

In the thermal-magnetic trip element circuit breaker, a magnetic element (electromagnet) is connected in series with the circuit load, and a bimetallic element is heated by the load current. With normal circuit current, the bimetallic element does not bend, and the magnetic element does not attract the trip bar, as shown in figure 2-21, view A.

If the temperature or current increases over a sustained period of time, the bimetallic element will bend, push the trip bar and release the latch. The circuit breaker will trip as shown in figure 2-21, view B.

If the current suddenly or rapidly increases enough, the magnetic element will attract the trip bar, release the latch, and the circuit breaker will trip, as shown in figure 2-21, view C. (This circuit breaker has tripped even though the thermal element has not had time to react to the increased current.)

Q30. *What are the five main components of a circuit breaker?*

Q31. *What are the three types of circuit breaker trip elements?*

Q32. *How does each type of trip element react to an overload?*

TRIP-FREE/NONTRIP-FREE CIRCUIT BREAKERS

Circuit breakers are classified as being trip free or nontrip free. A trip-free circuit breaker is a circuit breaker that will trip (open) even if the operating mechanism (ON-OFF switch) is held in the ON position. A nontrip-free circuit breaker can be reset and/or held ON even if an overload or excessive heat condition is present. In other words, a nontrip-free circuit breaker can be bypassed by holding the operating mechanism ON.

Trip-free circuit breakers are used on circuits that cannot tolerate overloads and on nonemergency circuits. Examples of these are precision or current sensitive circuits, nonemergency lighting circuits, and nonessential equipment circuits.Nontrip-free circuit breakers are used for circuits that are essential for operations. Examples of these circuits are emergency lighting, required control circuits, and essential equipment circuits.

TIME DELAY RATINGS

Circuit breakers, like fuses, are rated by the amount of time delay. In circuit breakers the ratings are instantaneous, short time delay, and longtime delay. The delay times of circuit breakers can be used to provide for SELECTIVE TRIPPING.

Selective tripping is used to cause the circuit breaker closest to the faulty circuit to trip. This will remove power from the faulty circuit without affecting other, nonfaulty circuits. Figure 2-22 should help you understand selective tripping.

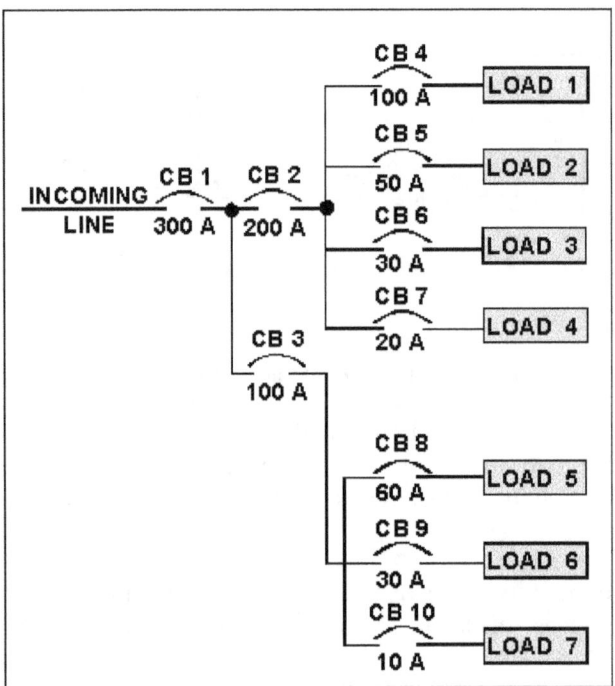

Figure 2-22.—Use of circuit breakers in a power distribution system.

Figure 2-22 shows a power distribution system using circuit breakers for protection. Circuit breaker 1 (CB1) has the entire current for all seven loads through it. CB2 feeds loads 1, 2, 3, and 4 (through CB4, CB5, CB6, and CB7), and CB3 feeds loads 5, 6, and 7 (through CB8, CB9, and CB10). If all the circuit breakers were rated with the same time delay, an overload on load 5 could cause CB1, CB3, and CB8 to trip. This would remove power from all seven loads, even though load 5 was the only circuit with an overload.

Selective tripping would have CB1 rated as long time delay, CB2 and CB3 rated as short time delay, and CB4 through CB10 rated as instantaneous. With this arrangement, if load 5 had an overload, only CB8 would trip. CB8 would remove the power from load 5 before CB1 or CB3 could react to the overload. In this way, only load 5 would be affected and the other circuits would continue to operate.

PHYSICAL TYPES OF CIRCUIT BREAKERS

All the circuit breakers presented so far in this chapter have been physically large, designed to control large amounts of power, and used a type of toggle operating mechanism. Not all circuit breakers are of this type. The circuit breaker in figure 2-23 is physically large and controls large amounts of power; but the operating mechanism is not a toggle. Except for the difference in the operating mechanism, this circuit breaker is identical to the circuit breakers already presented.

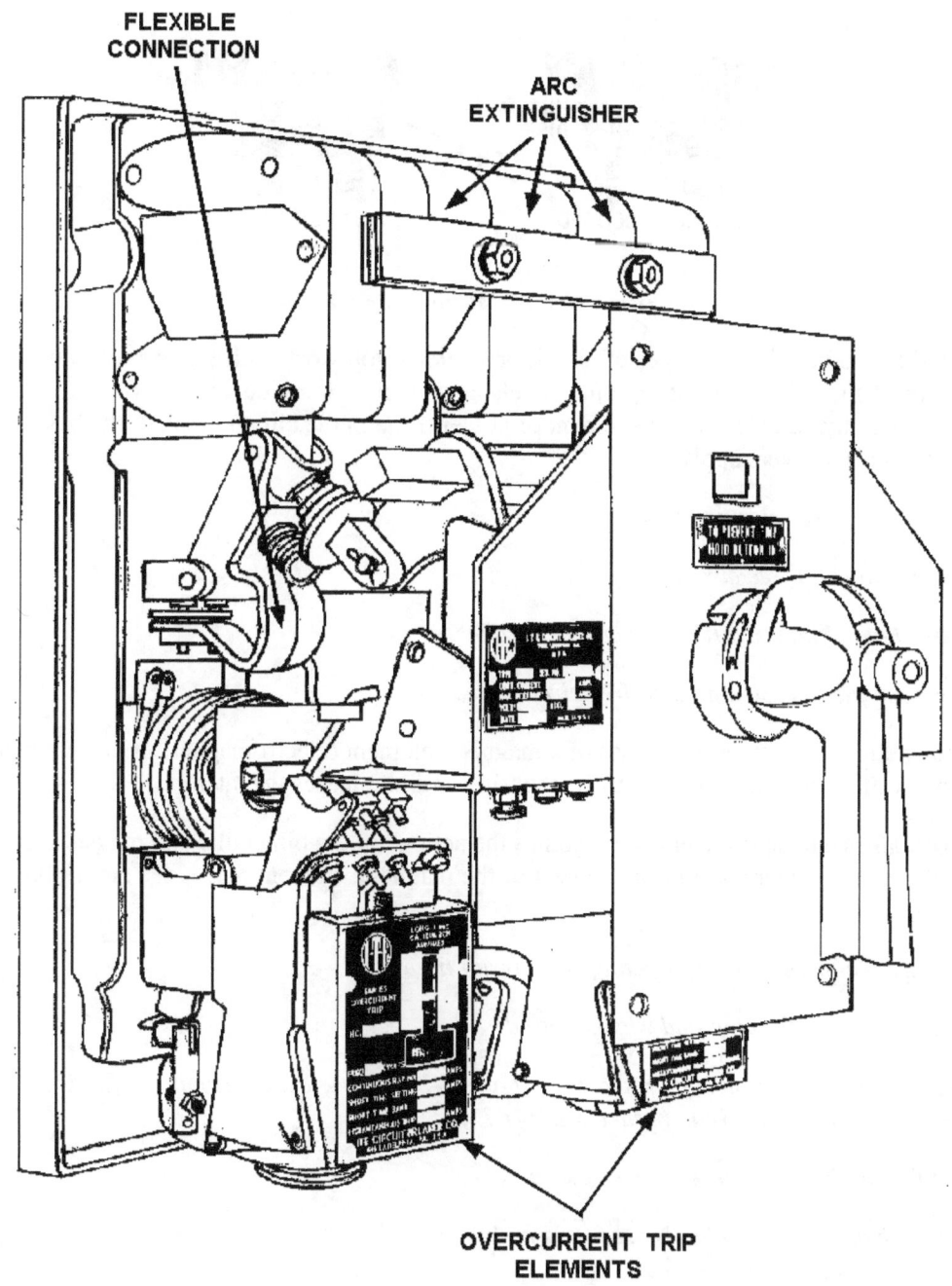

FLEXIBLE CONNECTION

ARC EXTINGUISHER

OVERCURRENT TRIP ELEMENTS

Figure 2-23.—Circuit breaker with an operating handle.

Circuit breakers used for low power protection, such as 28-volt dc, 30 amperes, can be physically small. With low power use, arc extinguishers are not required, and so are not used in the construction of these circuit breakers. Figure 2-24 shows a low power circuit breaker of the push-button or push-pull type. This circuit breaker has a thermal trip element (the bimetallic disk) and is nontrip-free. The push button is the operating mechanism of this circuit breaker.

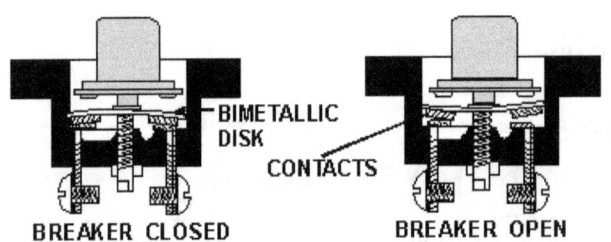

Figure 2-24.—Push-button circuit breaker.

You will find other physical types of circuit breakers as you work with electrical circuits. They are found in power distribution systems, lighting panels, and even on individual pieces of equipment. Regardless of the physical size and the amount of power through the circuit breaker, the basic operating principles of circuit breakers apply.

Q33. What is a trip-free circuit breaker?

Q34. What is a nontrip-free circuit breaker?

Q35. Where should you use a trip-free circuit breaker?

Q36. Where should you use a nontrip-free circuit breaker?

The magnetic trip element makes use of a magnetic element (electromagnet). If current reaches a preset quantity, the magnetic element attracts the trip bar and releases the latch.

The thermal-magnetic trip element combines the actions of the bimetallic and magnetic elements in a single trip element. If either the bimetal element or the magnetic element reacts, the circuit breaker will trip.

Q37. What are the three time delay ratings for circuit breakers?

Q38. What is selective tripping and why is it used?

Q39. If the power distribution system shown in figure 2-22 uses selective tripping, what is the time delay rating for each of the circuit breakers shown?

Q40. What factors are used to select a circuit breaker?

Q41. What type of circuit breaker is used on a multimeter?

CIRCUIT BREAKER MAINTENANCE

Circuit breakers require careful inspection and periodic cleaning. Before you attempt to work on circuit breakers, check the applicable technical manual carefully. When you work on shipboard circuit breakers, the approval of the electrical or engineering officer must be obtained before starting work. Be certain to remove all power to the circuit breaker before you work on it. Tag the switch that removes the power to the circuit breaker to ensure that power is not applied while you are working.

Once approval has been obtained, the incoming power has been removed, the switch tagged, and you have checked the technical manual, you may begin to check the circuit breaker. Manually operate the circuit breaker several times to be sure the operating mechanism works smoothly. Inspect the contacts for pitting caused by arcing or corrosion. If pitting is present, smooth the contacts with a fine file or number 00 sandpaper. Be certain the contacts make proper contact when the operating mechanism is ON.

Check the connections at the terminals to be certain the terminals and wiring are tight and free from corrosion. Check all mounting hardware for tightness and wear. Check all components for wear. Clean the circuit breaker completely.

When you have finished working on the circuit breaker, restore power and remove the tag from the switch that applies power to the circuit.

Q42. What steps are to be taken before beginning work on a circuit breaker?

Q43. What items are you to check when working on a circuit breaker?

SUMMARY

This chapter has provided the information to enable you to have a basic understanding of circuit protection devices. The following is a summary of the main points in this chapter.

CIRCUIT PROTECTION DEVICES are needed to protect personnel and circuits from hazardous conditions. The hazardous conditions can be caused by a direct short, excessive current, or excessive heat. Circuit protection devices are always connected in series with the circuit being protected.

A DIRECT SHORT is a condition in which some point in the circuit, where full system voltage is present, comes in direct contact with the ground or return side of the circuit.

EXCESSIVE CURRENT describes a condition that is not a direct short but in which circuit current increases beyond the designed current carrying ability of the circuit.

EXCESSIVE HEAT describes a condition in which the heat in or around a circuit increases to a higher than normal level.

FUSES and **CIRCUIT BREAKERS** are the two types of circuit protection devices discussed in this chapter.

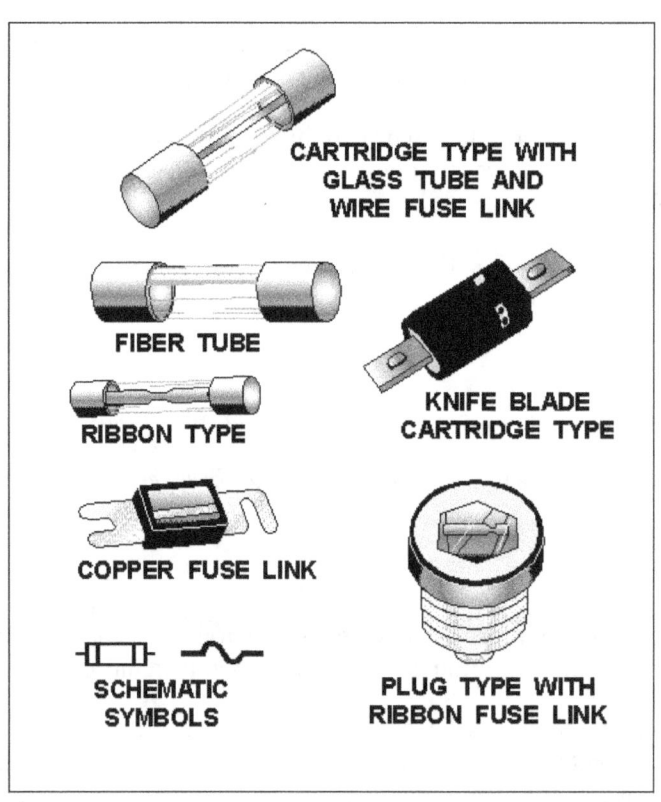

CARTRIDGE TYPE WITH GLASS TUBE AND WIRE FUSE LINK

FIBER TUBE

RIBBON TYPE

KNIFE BLADE CARTRIDGE TYPE

COPPER FUSE LINK

SCHEMATIC SYMBOLS

PLUG TYPE WITH RIBBON FUSE LINK

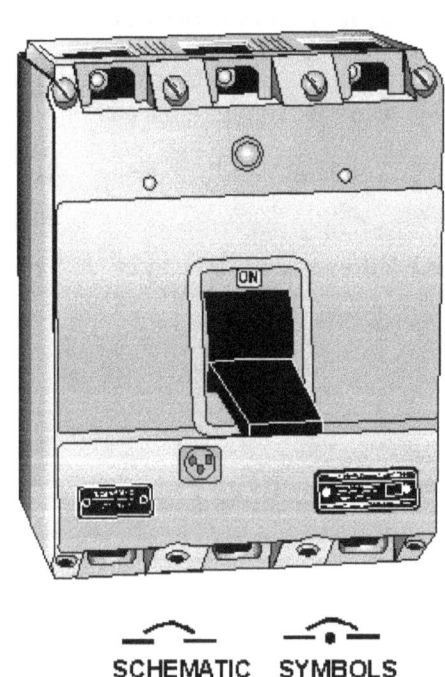

SCHEMATIC SYMBOLS

PLUG-TYPE FUSES are used in low-voltage, low-current circuits. This type fuse is rapidly being replaced by the circuit breaker.

CARTRIDGE FUSES are available in a wide range of physical sizes and voltage and current ratings. This type fuse is the most commonly used fuse.

The **CURRENT RATING** of a fuse is a value expressed in amperes that represents the amount of current the fuse will allow to flow without opening.

The **VOLTAGE RATING** of a fuse indicates the ability of the fuse to quickly extinguish the arc after the fuse element melts and the maximum voltage the open fuse will block.

The **TIME DELAY RATING** of a fuse indicates the relationship between the current through the fuse and the time it takes for the fuse to open. The three time delay ratings for fuses are DELAY, STANDARD, and FAST.

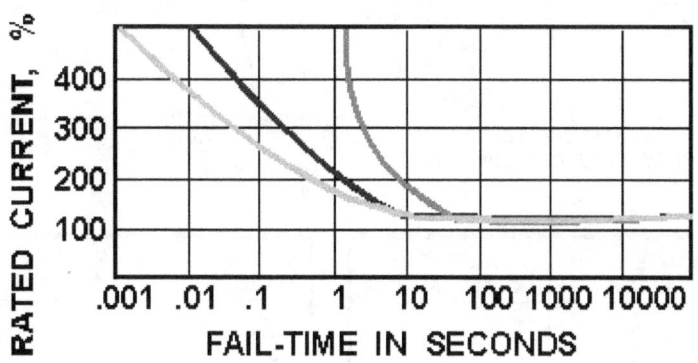

DELAY FUSES allow surge currents without opening. They are used to protect motors, solenoids, and transformers.

STANDARD FUSES have neither a time delay nor a fast acting characteristic. They are used in automobiles, lighting circuits and electrical power circuits.

FAST FUSES open very quickly with any current above the current rating of the fuse. They are used to protect delicate instruments or semiconductor devices.

The **OLD MILITARY FUSE DESIGNATION** is a system of fuse identification that uses coding to represent the current, voltage, and time-delay rating of the fuse. New fuses purchased by the Navy will no longer use this designation.

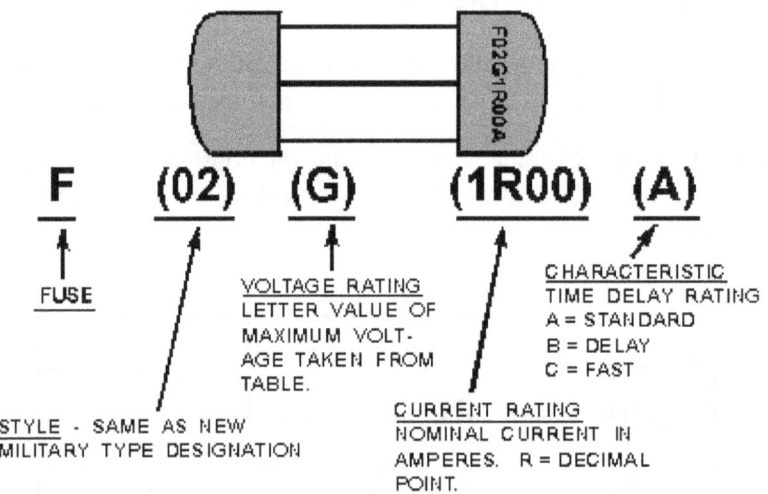

VOLTAGE CODE LETTER	VOLTAGE (VOLTS OR LESS)
A	32
B	52
C	90
D	125
G	250
H	500
J	1,000
K	2,500
N	5,000
P	10,000

VOLTAGE CODE (OLD)

CURRENT CODE	CURRENT (AMPERES)
R.002	.002 = 1/500
R.005	.005 = 1/200
R.010	.010 = 1/100
R.031	.031 = 1/32
R.750	.750 = 3/4
1R.50	1.500 = 1 1/2

CURRENT CODE (OLD)

The **NEW MILITARY FUSE DESIGNATION** is the system used to identify fuses purchased by the Navy at the present time. The coding of current and voltage ratings has been replaced with direct printing of these ratings.

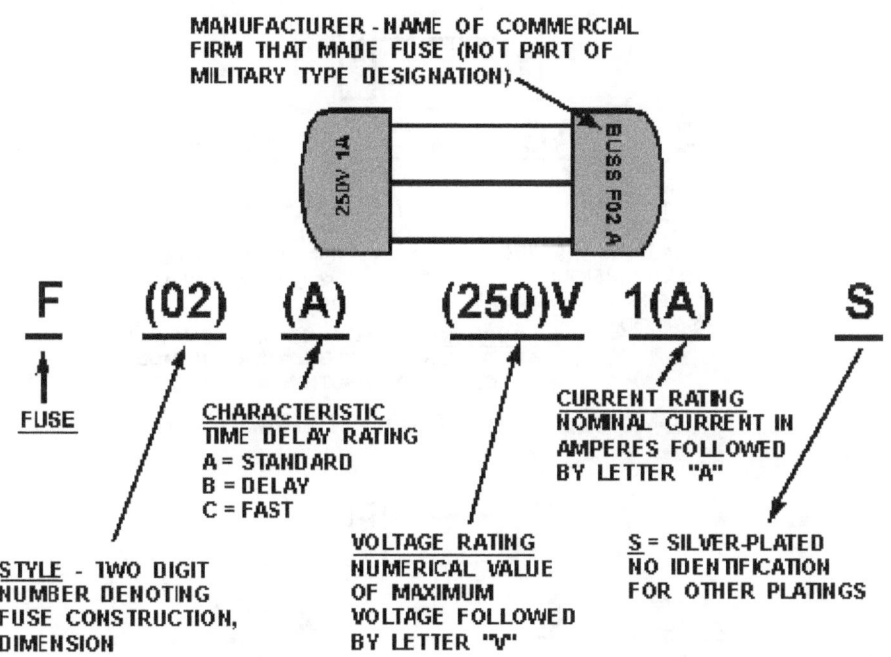

MANUFACTURER - NAME OF COMMERCIAL
FIRM THAT MADE FUSE (NOT PART OF
MILITARY TYPE DESIGNATION)

F (02) (A) (250)V 1(A) S

FUSE

CHARACTERISTIC
TIME DELAY RATING
A = STANDARD
B = DELAY
C = FAST

CURRENT RATING
NOMINAL CURRENT IN
AMPERES FOLLOWED
BY LETTER "A"

STYLE - TWO DIGIT
NUMBER DENOTING
FUSE CONSTRUCTION,
DIMENSION

VOLTAGE RATING
NUMERICAL VALUE
OF MAXIMUM
VOLTAGE FOLLOWED
BY LETTER "V"

S = SILVER-PLATED
NO IDENTIFICATION
FOR OTHER PLATINGS

The **OLD COMMERCIAL FUSE DESIGNATION** was used by the fuse manufacturers to identify fuses. The current and voltage ratings are printed on the fuse, but the time delay rating is contained in the style coding of the fuse.

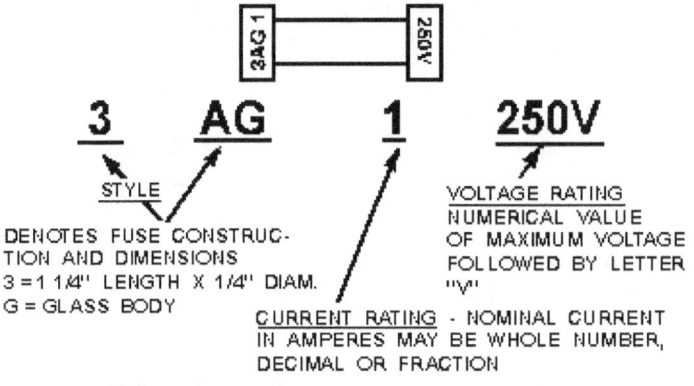

(A) OLD COMMERCIAL DESIGNATIONS

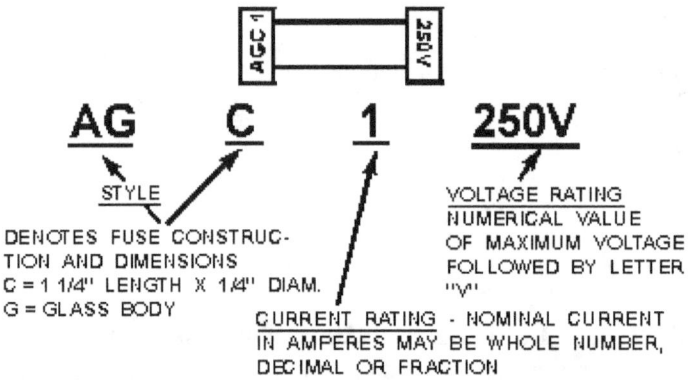

(B) NEW COMMERCIAL DESIGNATIONS

The **NEW COMMERCIAL FUSE DESIGNATION** is currently used by fuse manufacturers to identify fuses. It is similar to the old commercial fuse designation with the difference being in the style coding portion of the designation.

FUSEHOLDERS are used to allow easy replacement of fuses in a circuit.

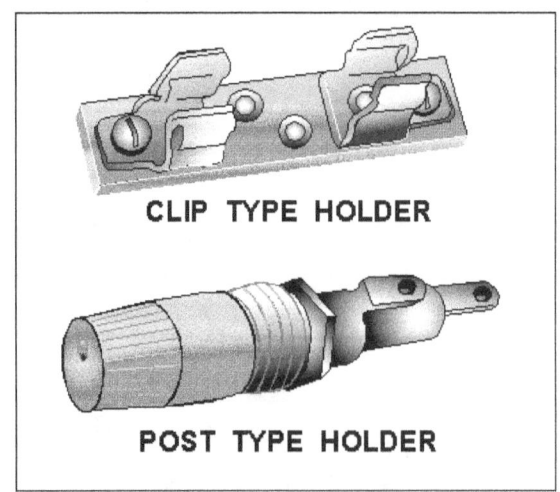

CLIP TYPE HOLDER

POST TYPE HOLDER

The **CLIP-TYPE** has clips to connect the ferrules or knifeblades of the fuse to the circuit. The POST-TYPE is an enclosed fuseholder. The center connection of the post type should be connected to the power source and the outside connector should be connected to the load.

OPEN FUSES can be found by VISUAL INSPECTION, FUSE INDICATORS, or by the use of a METER. The following SAFETY PRECAUTIONS should be observed when checking a fuse:

- Turn the power off and discharge the circuit before removing a FUSE.

- Use a fusepuller when you remove a fuse from clip-type fuseholders.

- When you check a fuse with a voltmeter, be careful to avoid shocks and short circuits.

- When you use an ohmmeter to check fuses with low current ratings, be careful to avoid opening the fuse by excessive current.

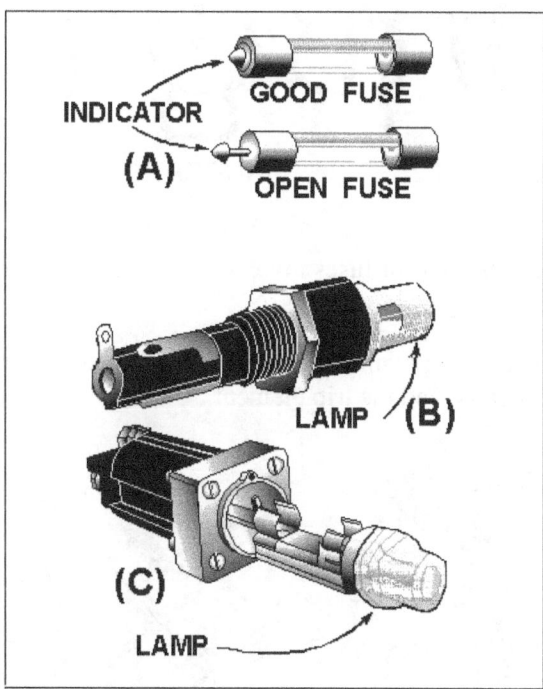

REPLACEMENT FUSES must be of the proper type. Check the technical manual parts list to find the identification of the proper fuse. If a substitute fuse must be used, the following guidelines should be followed:

- Never use a fuse with a higher current rating, a lower voltage rating, or a slower time delay rating than the specified fuse.

- The best substitution fuse is a fuse with the same current and time delay ratings and a higher voltage rating.

- If a lower current rating, or a lower time delay rating is used, the fuse may open under normal circuit conditions. Substitute fuses must have the same style (physical dimensions) as the specified fuse.

PROPER FIT between the fuse and fuseholder is essential. If the clips on clip-type fuseholders are sprung, the clips should be reformed, or clip clamps should be used. Any corrosion on fuses or fuseholders must be removed with fine sandpaper.

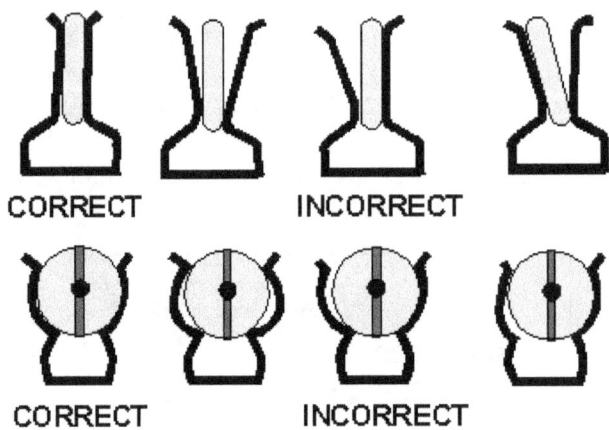

PREVENTIVE MAINTENANCE of fuses involves checking for the proper fuse, corrosion, proper fit, and open fuses; and correcting any discrepancies.

CIRCUIT BREAKERS have five main components: The frame, the operating mechanism, the arc extinguisher, the terminal connectors, and the trip element.

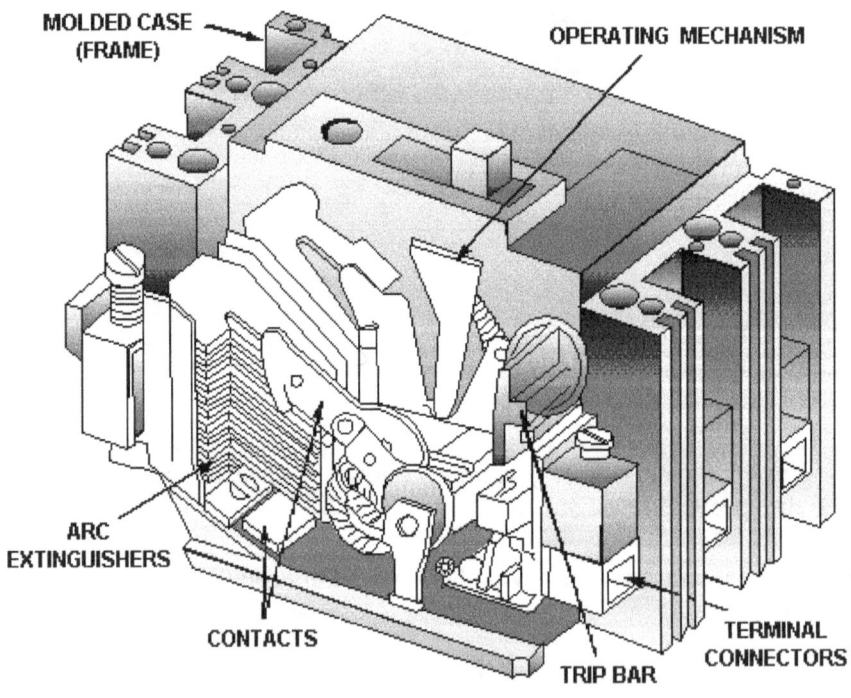

A **THERMAL TRIP ELEMENT** uses a bimetallic element that is heated by load current and bends due to this heating. If current (or temperature) increases above normal, the bimetallic element bends to push against a trip bar and opens the circuit.

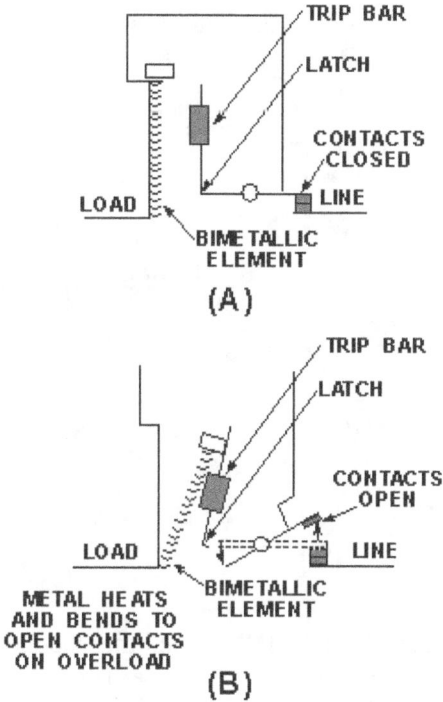

(A)

(B)

A **MAGNETIC TRIP ELEMENT** uses an electromagnet in series with the load current to attract the trip bar and open the circuit if excessive current is present.

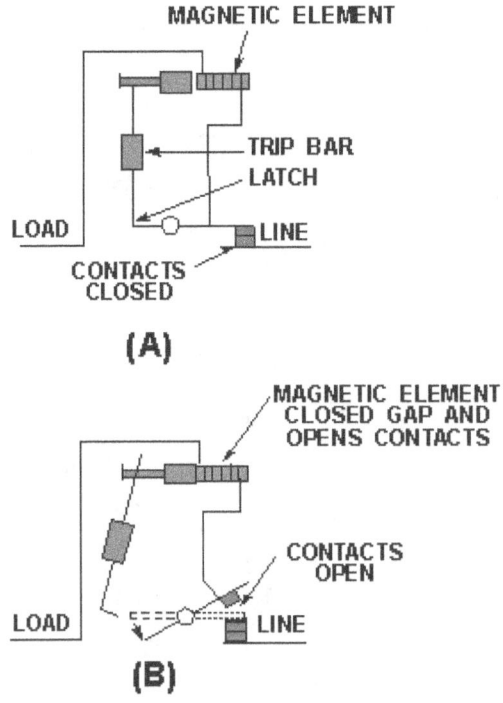

A **THERMAL-MAGNETIC TRIP ELEMENT** combines the thermal and magnetic trip elements into a single unit. A TRIP-FREE circuit breaker will trip (open) even if the operating mechanism is held in the ON position. A TRIP-FREE circuit breaker would be used on non-essential circuits.

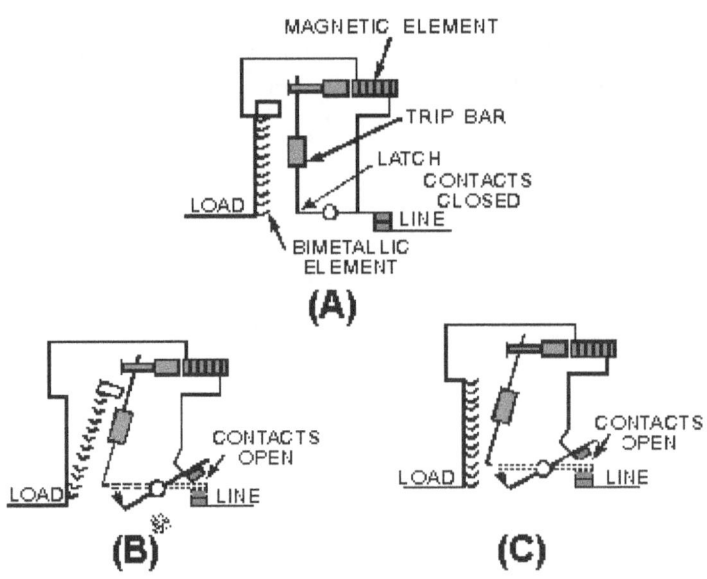

A **NONTRIP-FREE** circuit breaker can be bypassed by holding the operating mechanism ON. A NONTRIP-FREE circuit breaker would be used for emergency or essential equipment circuits.

The **TIME DELAY RATINGS** of circuit breakers are INSTANTANEOUS, SHORT TIME DELAY, and LONG TIME DELAY.

SELECTIVE TRIPPING is used to cause the circuit breaker closest to the faulty circuit to trip, isolating the faulty circuit without affecting other nonfaulty circuits. This is accomplished by using an instantaneous circuit breaker close to the load, a short time delay circuit breaker at the next junction, and a long time delay circuit breaker at the main junction box.

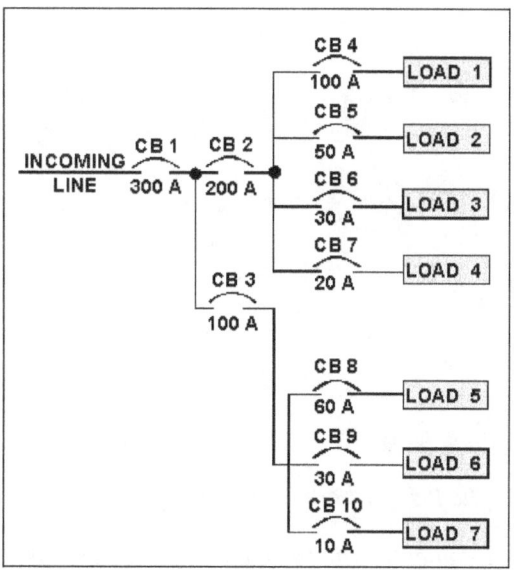

The **FACTORS** used to select a circuit breaker are the power requirements of the circuit and the physical space available.

When **WORKING ON CIRCUIT BREAKERS**, the following items should be done BEFORE working on the circuit breaker: Check the applicable technical manual, obtain the approval of the electrical or engineering officer (for shipboard circuit breakers), remove power from the circuit breaker, and tag the switch that removes power from the circuit breaker. The following items should be checked and discrepancies corrected when working on circuit breakers: Check the operating mechanism for smooth operation, check the contacts for pitting, check the terminals for tightness and corrosion, check the mounting hardware for tightness and wear, check all components for wear, and check the entire circuit breaker for cleanliness.

ANSWERS TO QUESTIONS Q1. THROUGH Q43.

A1. To protect people and circuits from possible hazardous conditions.

A2. A direct short, excessive current, and excessive heat.

A3. A condition in which some point in the circuit where full system voltage is present comes in contact with the ground or return side of the circuit.

A4. A condition that is not a direct short but in which circuit current increases beyond the designed current carrying ability of the circuit.

A5. A condition in which the heat in or around the circuit increases to a higher than normal level

A6. *In series, so total current will be stopped when the device opens.*

A7. *Fuses and circuit breakers.*

A8.

     a. *circuit breaker*

     b. *fuse.*

A9.

     a. *cartridge*

     b. *plug*

     c. *plug*

     d. *cartridge.*

A10. *A, C.*

A11. *Current, voltage, and time delay.*

A12. *The amount of current the fuse will allow without opening.*

A13. *The ability of the fuse to quickly extinguish the arc after the fuse element melts and the maximum voltage that cannot jump across the gap of the fuse after the fuse opens.*

A14. *Delay, standard, and fast.*

A15. *Delay-Motors, solenoids, or transformers. Standard-Automobiles, lighting or electrical power circuits. Fast-Delicate instruments or semiconductor devices.*

A16.

     a. *125 volts or less, 1.5 amperes, delay*

     b. *250 volts or less, 1/8 ampere standard*

A17.

     a. *125 volts or less, 1/16 ampere*

     b. *250 volts or less, .15 ampere*

A18. *F05B32V20A.*

A19.

     a. *Post-type fuseholder*

     b. *Clip-type fuseholder*

A20.

    a.   Center connector

    b.   Outside connector

A21. Visual inspection, indicators, and using a meter.

A22. Put it back in the circuit. A good fuse will have zero ohms of resistance.

A23. The ohmmeter causes more than 1/500 ampere through the fuse when you check the fuse, thus it opens the fuse.

A24. Use a resistor in series with the fuse when you check it with the ohmmeter.

A25. Turn the power off and discharge the circuit before you remove fuses. Use a fuse puller (an insulated tool) when you remove fuses front clip-type fuse holders. When you check fuses with a voltmeter, be careful to avoid shocks and short circuits.

A26.

    a.   Not acceptable-wrong style

    b.   Substitute #3-smaller current rating

    c.   Substitute #1-identical, except higher voltage rating

    d.   Not acceptable-lower voltage rating

    e.   Direct replacement

    f.   Not acceptable-higher current rating

    g.   Substitute #2-Faster time delay rating

A27. Check for the proper type of replacement fuse and proper fit.

A28. Be sure the power is off in the circuit and the circuit is discharged before replacing a fuse. Use an identical replacement fuse if possible. Remove any corrosion from the fuseholders before replacing the fuses.

A29. Improper fuse, corrosion, improper fit, and open fuse.

A30. Frame, operating mechanism, arc extinguishers, terminal connectors, and trip element.

A31. Thermal, magnetic, and thermal-magnetic.

A32. The thermal trip element makes use of a bimetallic element that bends with an increase in temperature or current. The bending causes the trip bar to be moved releasing the latch.

A33. A circuit breaker that will trip even if the operating mechanism is held ON.

A34. A circuit breaker that can be overridden if the operating mechanism is held ON.

A35. In current sensitive or nonemergency systems.

A36. *In emergency or essential circuits.*

A37. *Instantaneous, short time delay, and long time delay.*

A38. *It is the use of time delay ratings to cause the circuit breaker closest to the faulty circuit to trip. This isolates the faulty circuit without affecting other circuits.*

A39. *CB1-long time delay; CB2, CB3-short time delay; CB4 through CB10-instantaneous.*

A40. *The power requirements of the circuit and the physical space available.*

A41. *A push button or push-pull circuit breaker (small size, low power).*

A42. *Check the applicable technical manual, obtain the approval of the electrical or engineering officer (for shipboard circuit breakers), remove power from the circuit breaker, and tag the switch that supplies power to the circuit breaker.*

A43. *Check the operating mechanism for smooth operation, check the contacts for pitting, check the terminals for tightness and corrosion, check the mounting hardware for tightness and wear, check all components for wear, and check the entire circuit breaker for cleanliness.*

www.ingramcontent.com/pod-product-compliance
Lightning Source LLC
Chambersburg PA
CBHW080618180526
45168CB00007B/2966